Entangled Magazin[e]

MW00987642

Synthetic Biocomputing & Smart Vaccines

Table of Contents

During the past twenty years, the field of synthetic biology has experienced a parabolic acceleration of development and growth. Within it, new types of cellular sensors have been established for in vivo (in the cell, body) applications. These biological sensors (biosensors) are now paired to electronic computer chips, and transplanted or injected in tissues for monitoring, and computations using logic-gates (YES, NOT, INHIBIT, NOR, NAND and OR), and for targeted drug delivery applications such as vaccines.

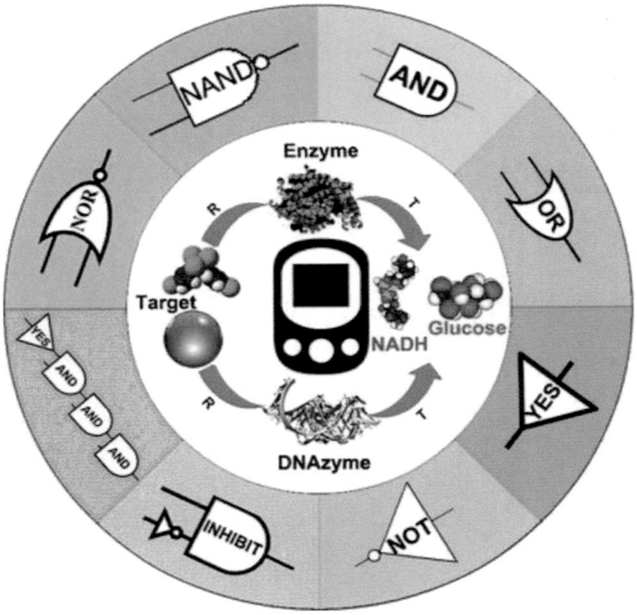

Glucose Meter a Logic-Gate Responsive Device

These accelerated advancements in synthetic biology and thus, biocomputation tools have produced new genetic circuits operating as logic functions. Digital circuits in cells are based on RNA regulators (riboswitches, toehold switches, etc.) and recombinases.

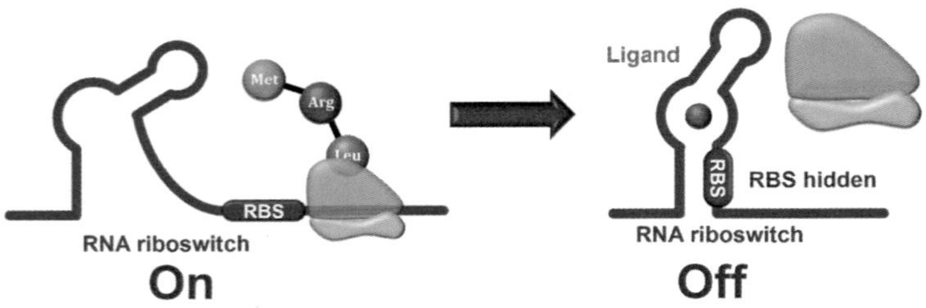

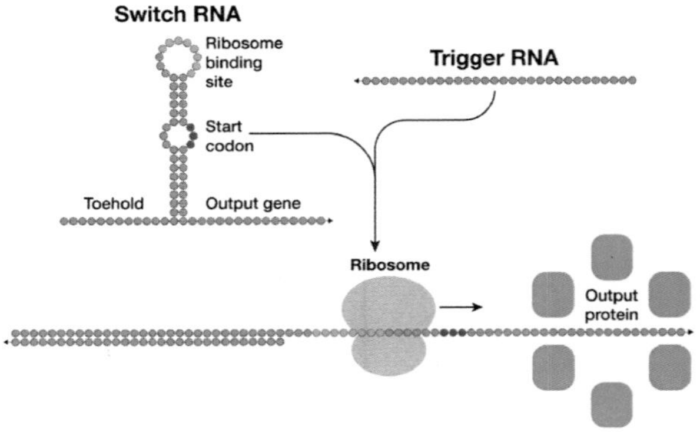

Within synthetic biology, research has focused upon the construction of circuits of biological components such as genes and the proteins they express. Circuits receive input signals from inside or outside the body. Encoded within circuits are so-called "rules", mimicking those found in conventional, classical computing based upon electronic logic circuits. In synthetic biology, logic circuits are derived from genes and the control of their activities, or the silencing of them.

These are both digital and analog circuits responding to a variety of inputs. Genetic circuits process digital-to-analog, analog-to-digital conversions, while counting, comparing and processing quantities of inputs.

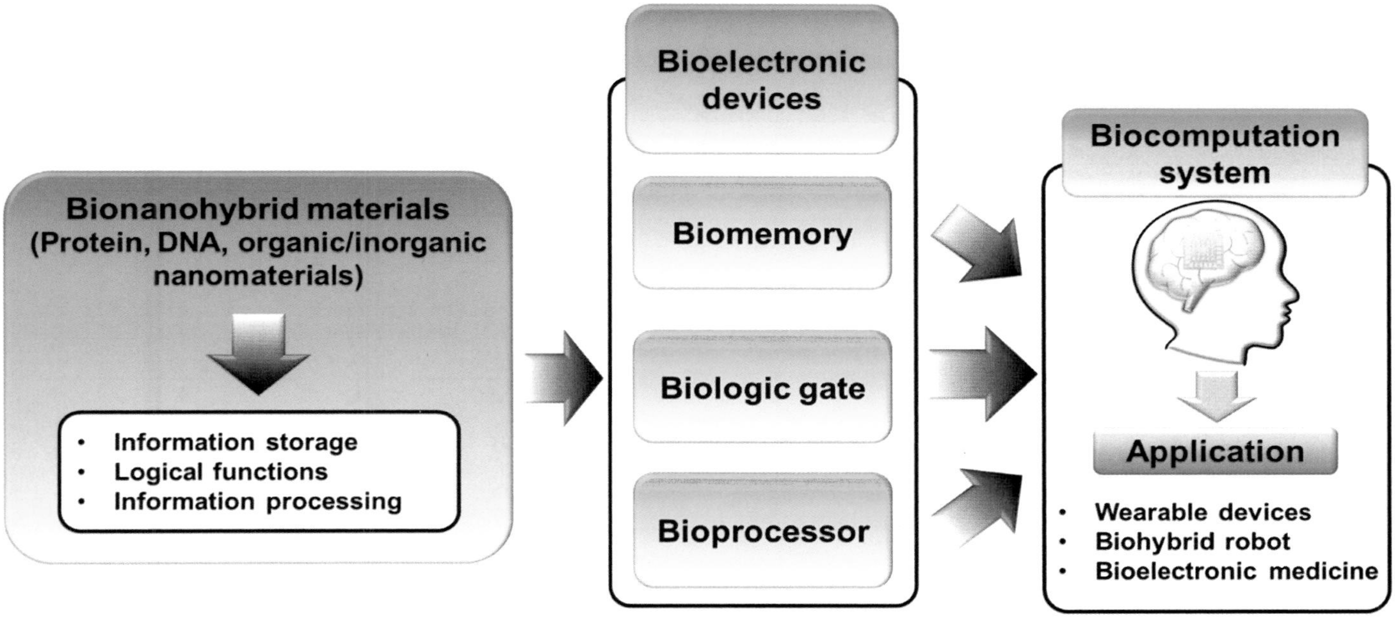

The naturally-derived biosensors within our bodies employ elements such as nucleic acids (DNA, RNA), proteins, enzymes, and antibodies from our innate and adaptive immune systems, as recognition layers. These layers then produce an output signal as found in standard electronic logic circuits.

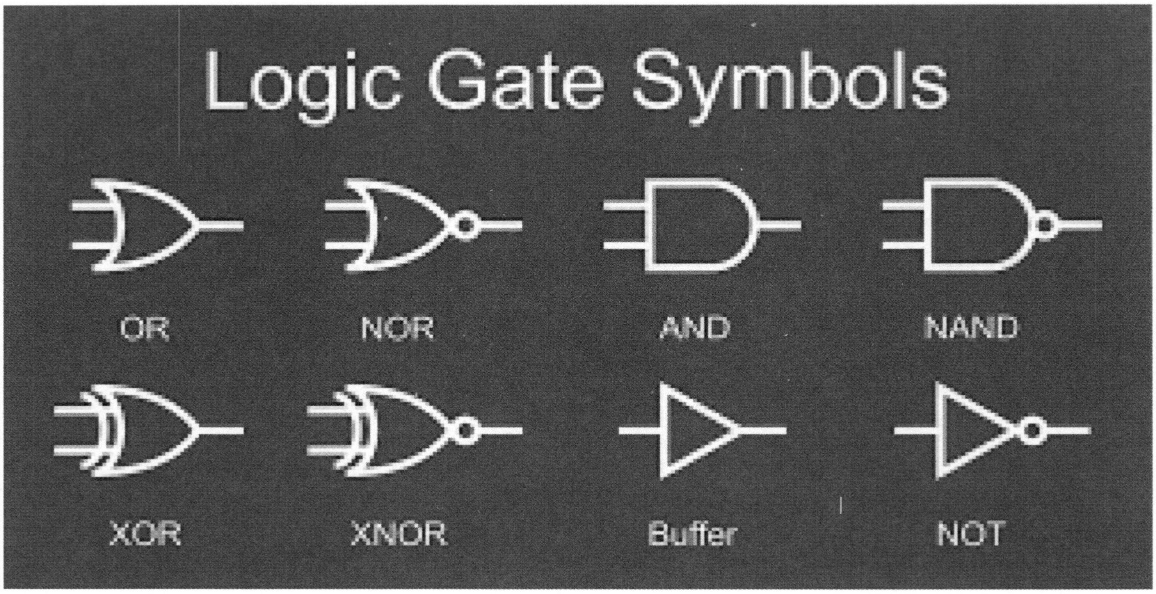
Logic Gate Symbols
OR
NOR
AND
NAND
XOR
XNOR
Buffer
NOT

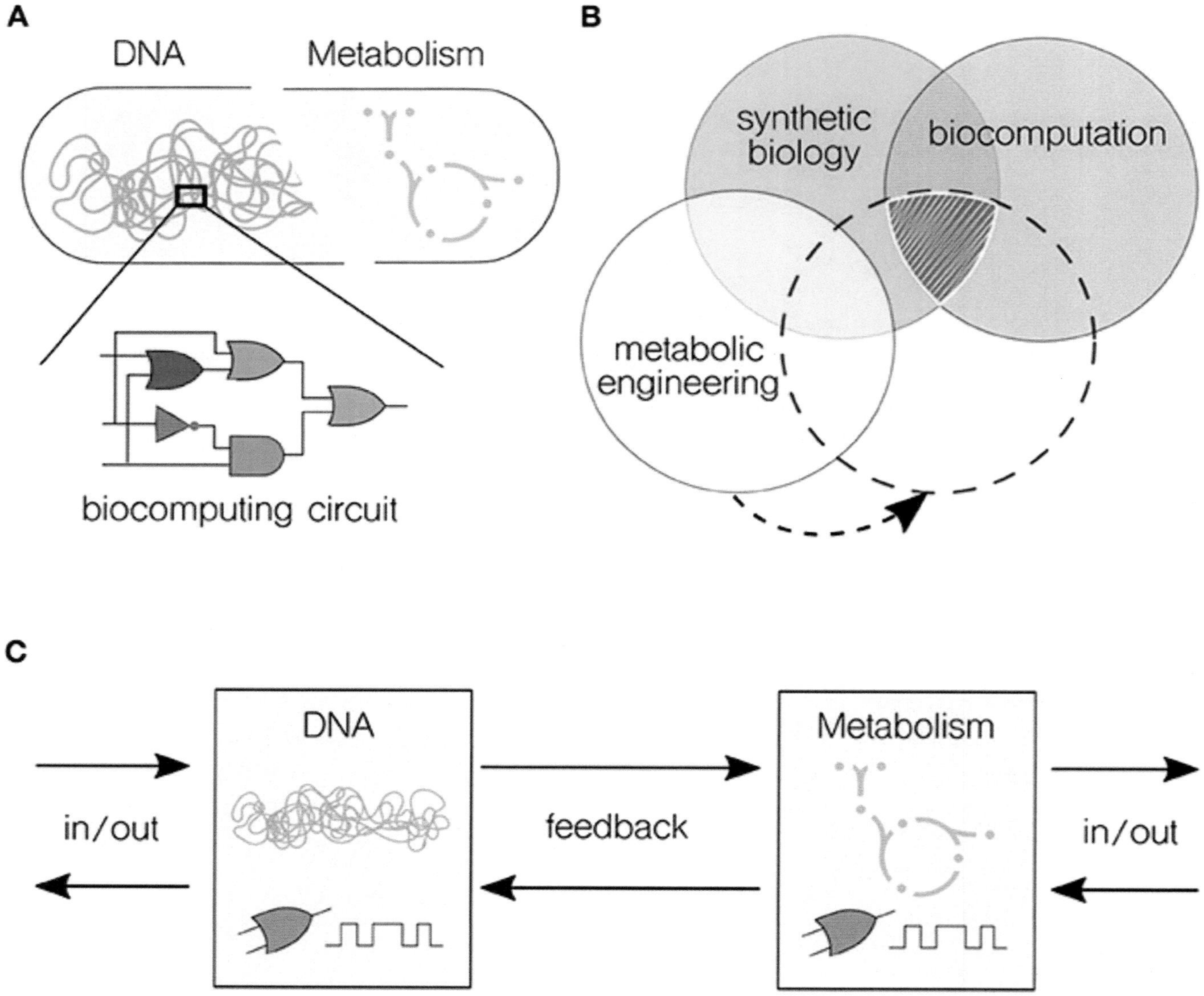
A
DNA
Metabolism
biocomputing circuit
B
synthetic biology
biocomputation
metabolic engineering
C
DNA
Metabolism
in/out
feedback
in/out

Over twenty years ago, the first synthetic gene circuits were developed in E. coli. Biologists have produced standardized genetic pieces and devices that are then arranged into modules and systems for reprogramming living organisms. The field now is more aptly viewed as bioengineering.

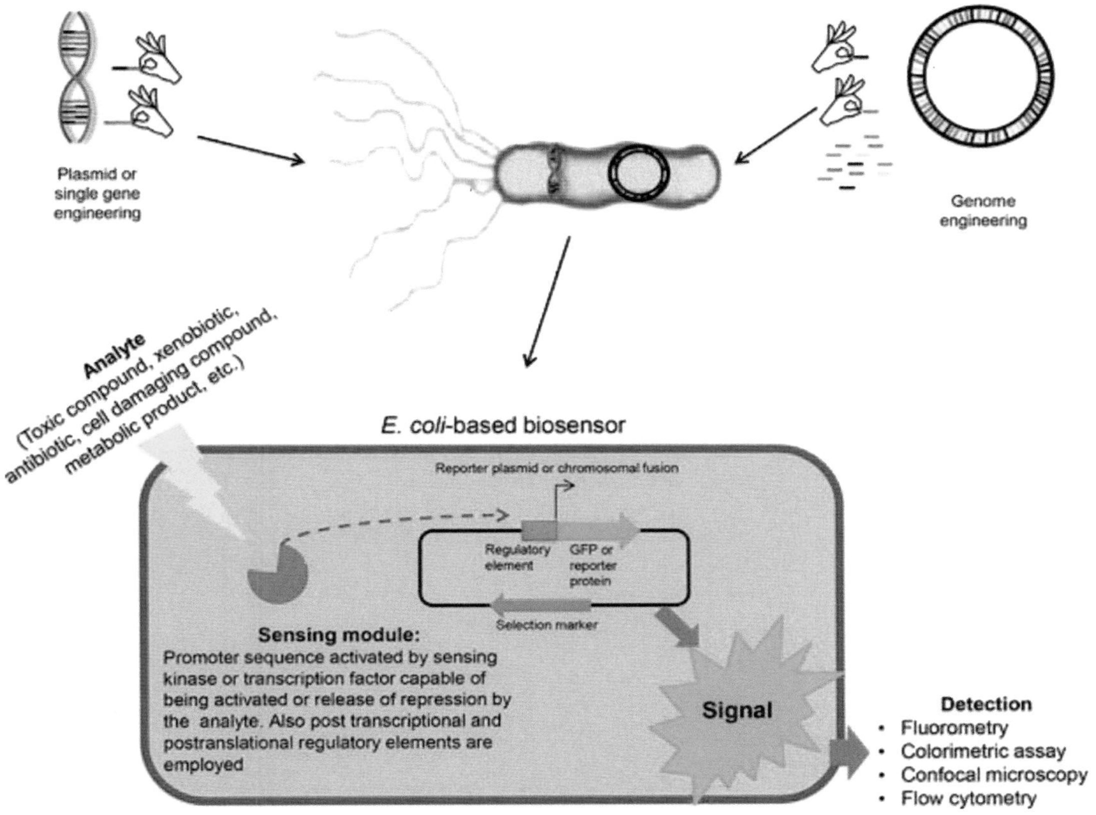

Bioengineering gave rise to biocomputing using molecular parts for the hardware. Biological systems have what are referred to as hard-coded, pre-defined "rules". Functional devices are built from molecular tools and biological motifs. This began with the small molecules of genes, arranging (engineering) them into designed genetic circuits patterned after electronic functions known as logic gates.

There are two primary categories working together in synthetic biocomputing. First, the genetic parts, the hardware and second, the cellular metabolism. Genetic components work in conjuction with the remaining metabolic cellular machinery. The result is whole-cell biocomputing utilizing both the transcriptional and the metabolic circuits. This increases the volume and type of information processing, and the implemenation of control mechanisms.

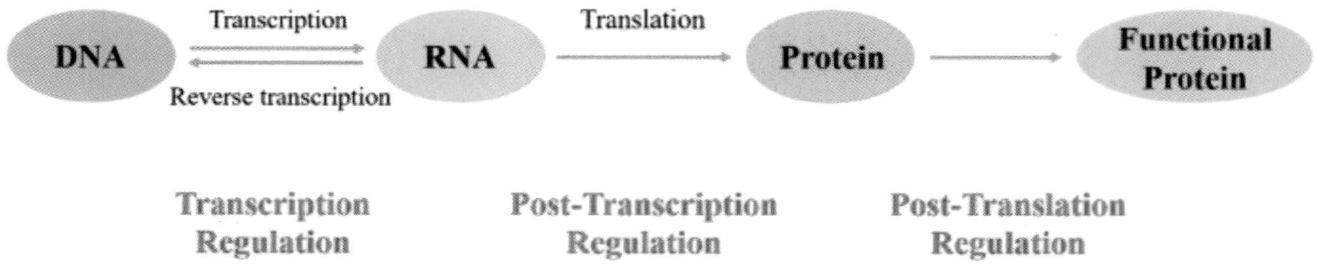

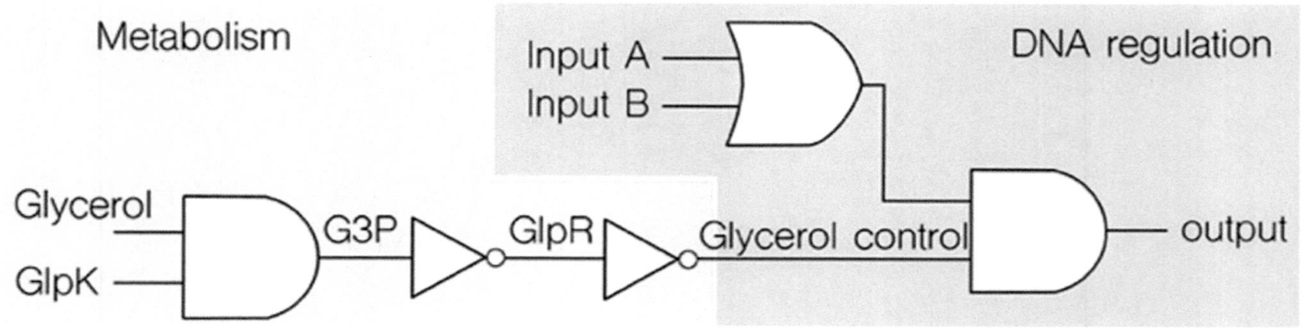

Within cells, genetic and metabolic systems work together. There is a concept known as heterotic computing (i.e., the coordination between different types of computing), which is inherent to biological systems. Metabolic networks operate as conserved principles that can be implemented to repurpose biochemical nodes.

Within the present context of developing vaccines based upon DNA/RNA, RNA is now encoded with synthetic gene circuits. The RNA is regulated using programmable antigen/adjuvant circuits. These are designed so as not to stimulate the innate or adaptive immune system responses in such a manner that works against the regulatory components of these circuits. At least that is the stated and published goal of the vaccine developers.

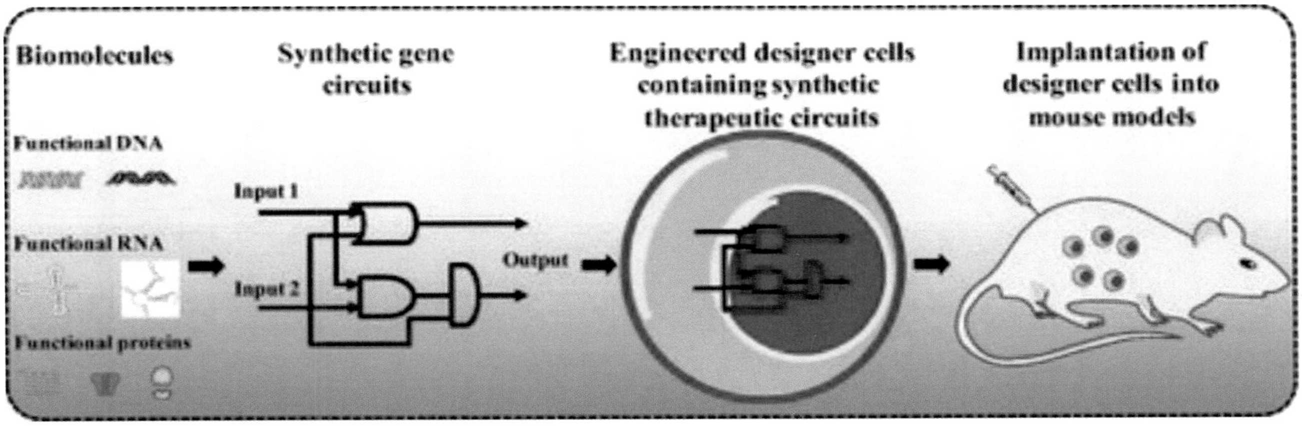

The postinoculation reports indicate otherwise, specifically with respect to the more severe side effects such as anaphylaxis, Bell's palsy, sepsis, major organ failure and sudden deaths.

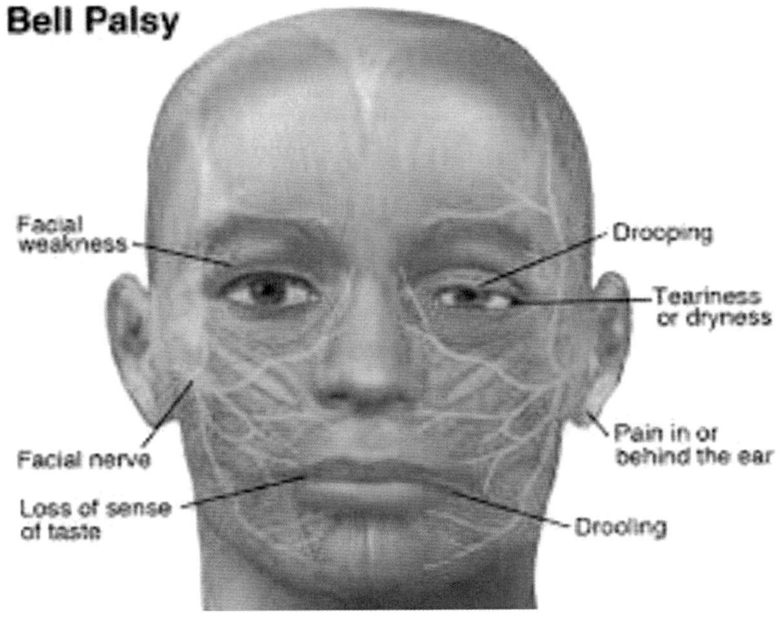

In silico mathematical design of RNA circuits results in their assembly in a top-down manner. This has completely changed the manner in which vaccines today are designed. In addition, programmable, individualized vaccines regulating the antigen/adjuvant expression levels are being produced.

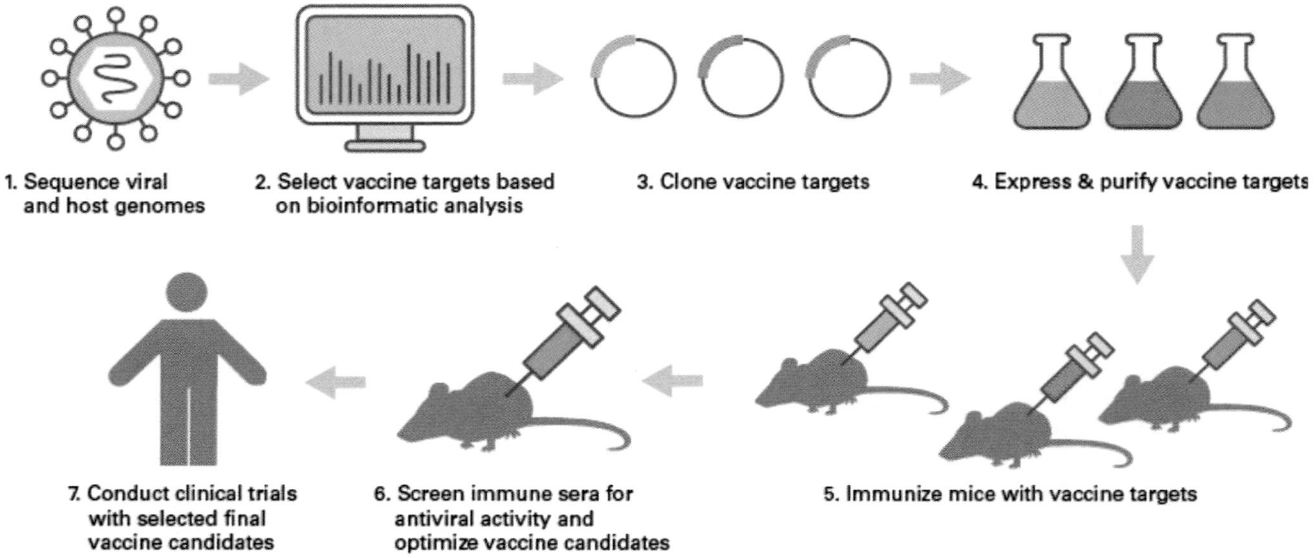

These are designed as alternatives to the more traditional vaccines employing an attenuated virus. This production and regulation of antigen/adjuvant expression is accomplished by delivering small molecule drugs as triggering mechanisms.

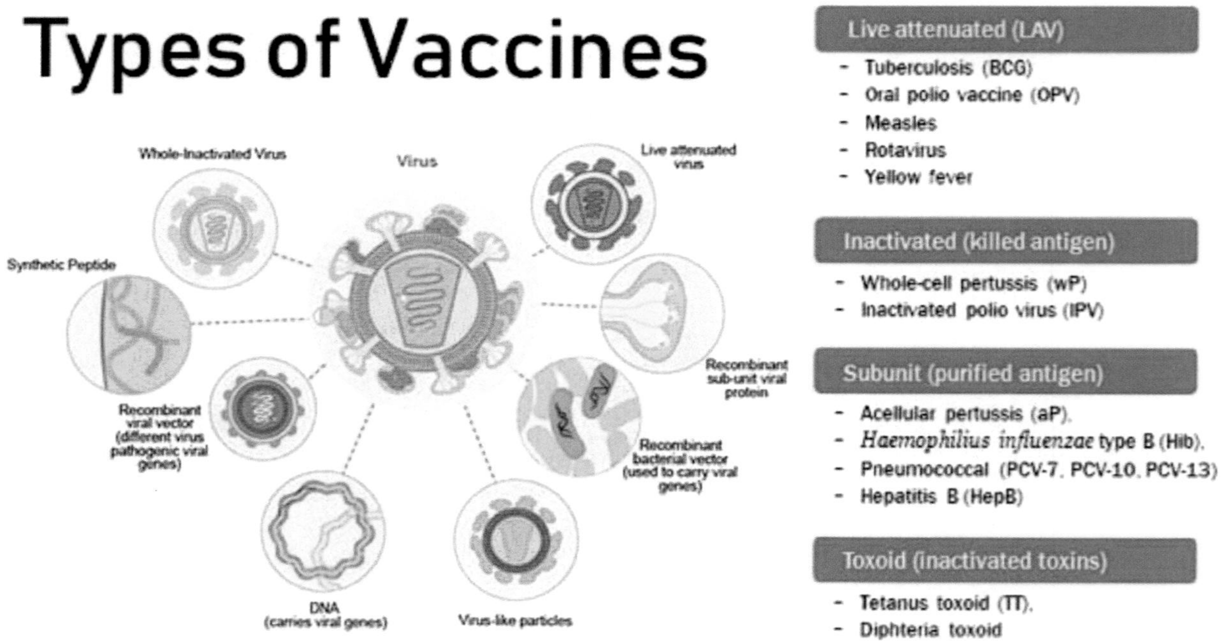

Genetic circuits work in a complementary fashion with small molecule drugs. These are oscillators, toggle switches and cascades designed in software, in silico, modeling logic-based networks. Working top-down from an examination of a gentic circuit, its high-level behavior of sensing-processing-actuation is used in the design process of constructing genetic circuits. While the physical action of these circuits is facilitated by the bottom-up assembly of the constituent biological parts. These logic circuits function as transcriptional, translational or posttranslational devices.

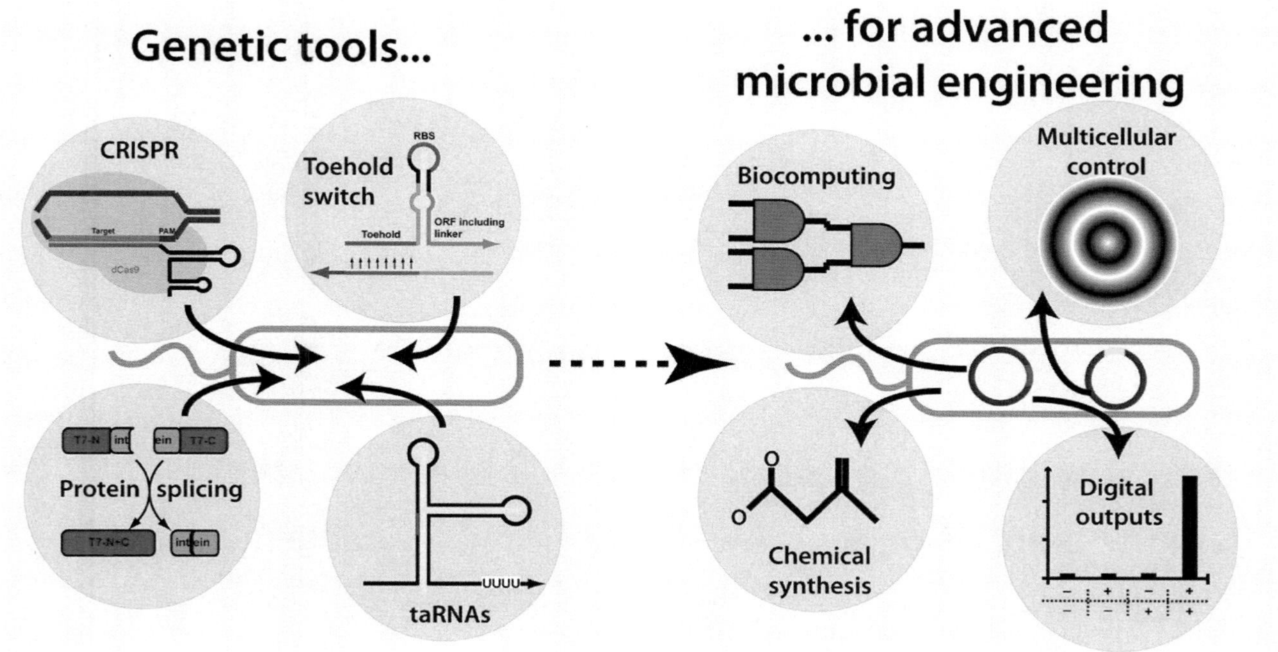

Over the past year, SARS-CoV-2 RNA-based synthetic vaccines have been developed and administered. These are vaccines with programmable adjuvant expression and prime-boost immunogen behavior. Meaning, these vaccines are designed and meant to be administered as one-shot injections, either by syringe and hypodermic needle, or microneedle array patches.

Within synthetic RNA circuits, singular inputs and outputs can be linked in producing complex modules. To accomplish this, the output of the first device must be compatible and convertable as the input for a second device. These can be integrated into systems of more advanced operations, beginning with in silico mathematical modeling of RNA circuits.

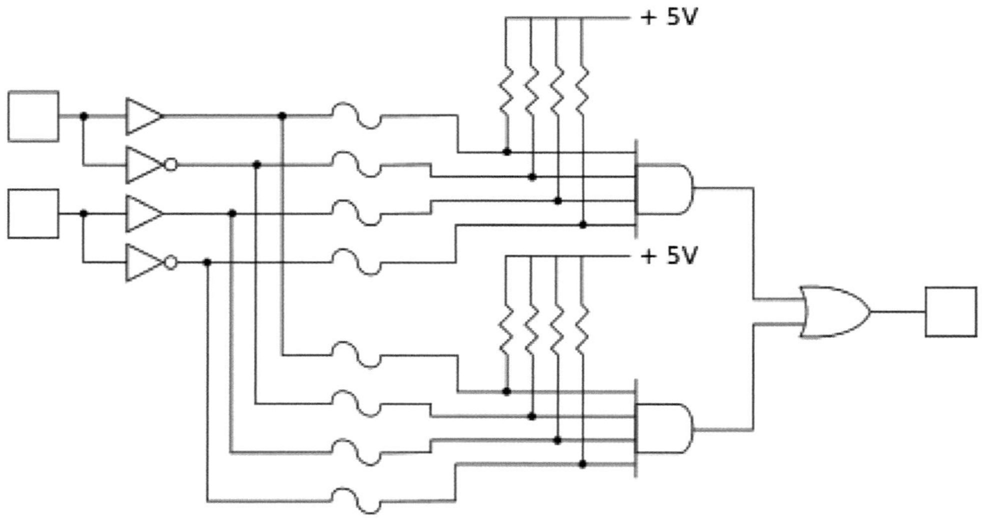

Simplified programmable logic device

A programmable logic device (PLD) is an electronic component used to build reconfigurable digital circuits. Unlike integrated circuits which consist of logic gates and have a fixed function, a PLD has an undefined function at the time of manufacture. Before the PLD can be used in a circuit it must be programmed (reconfigured) by using a specialized program. A device programmer is used to transfer the Boolean (a binary variable, having two possible values called "true" and "false") logic pattern into the programmable device.

Today, engineered cells operating as biosensors are delivered by inoculation to the human body. Using messenger RNA (mRNA) levels as inputs, biosensors combined with a computational module, are able to evaluate logic expressions. The relationship between computer science and natural, biological computing is often synergistic 1.

BIOSENSOR STRUCTURE

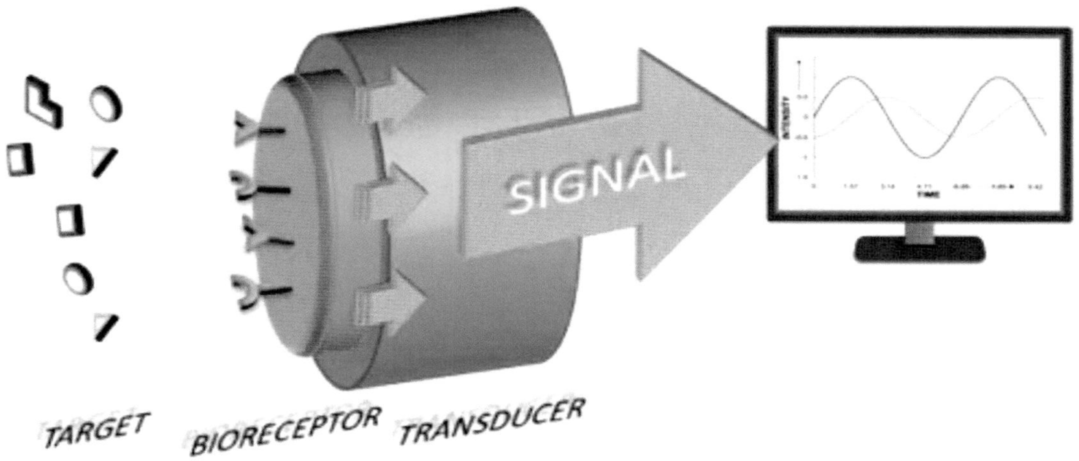

A biosensor is a small circuit that transduces a specific input signal into a prescribed output according to a pre-programmed input/output relation. When combined with an electronic nanoscale computation module running Boolean switching functions, a variety of computational models emerge leading to types of cellular features exceeding transistor-based logic circuits. Deployed on a global scale, DNA and mRNA-based vaccines are acting as micro-scale drug precursor production plants and miniturized biosensors 1.

In addition to drug production, nanoscale computational units utilizing both genetic materials and metabolic processes operate in tandem with arrays of biological sensors 1.

Bioretrosynthesis, a technique for synthesizing organic chemicals from inexpensive precursors and so-called 'evolved' enzymes, is an approach based on generic represenatations of reactions, known as reaction rules. These expand the design space of metabolic circuits by predicting new syntheic pathways connecting metabolism and gene expression. Artificial intelligence, known as machine-learning, is applied by such retrosynthesis algorithms to select the best candidate reactions 1.

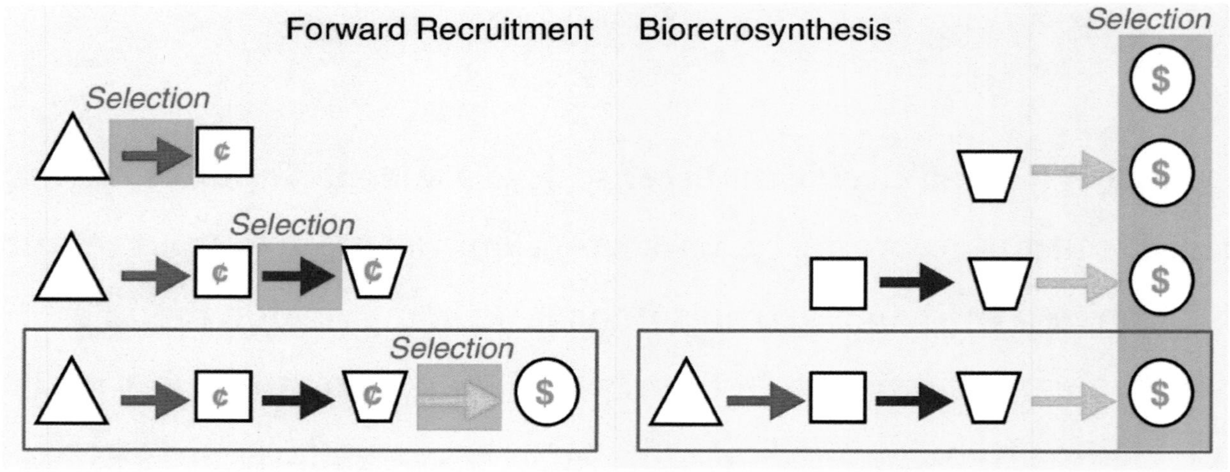

As in most computer architectures, the working relationship between gene expression and metabolism is akin to memory systems. For example, DNA is relatively stable, thus operating as long-term static memory. While metabolism is a short-term volatile memory system.

The current definition of digital computation is based on the abstract model defined by Alan Turing in the 1930s 2 and the John von Neumann architecture 3 used to implement the types of computations performed by the Turing Machine 1. Although Turing's model provides a framework for answering fundamental questions about computation, "...as soon as one leaves the comfort provided by the abundance of mathematical machinery used to describe digital computation, the world seems to be packed with pardoxes" 4.

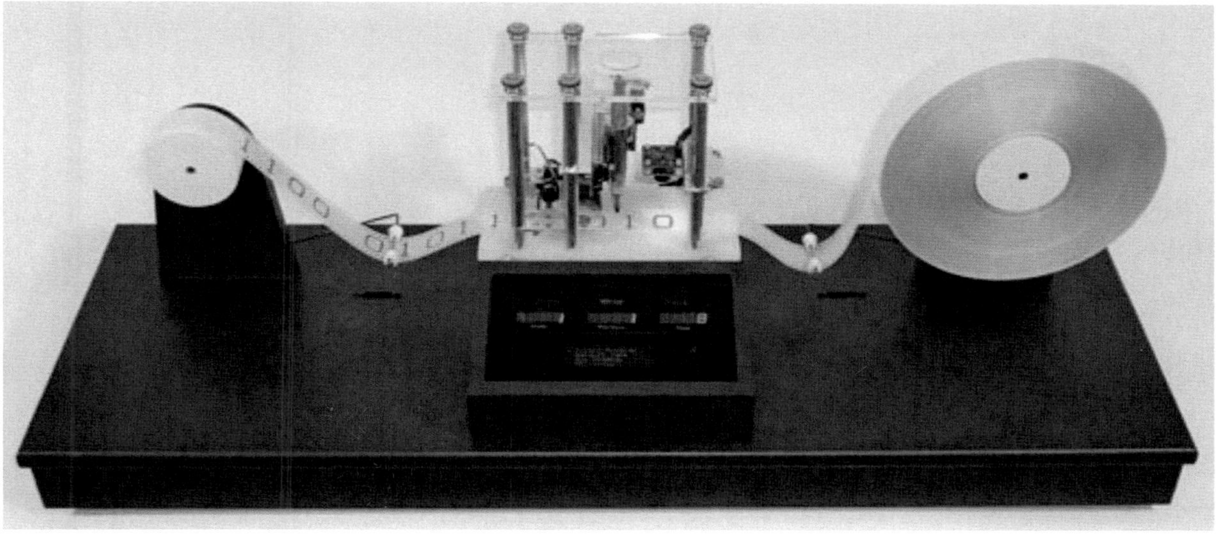

Turing Machine

Although genetic circuits may appear to behave digitally, it is only the collective behavior of a large number of inherently analog components that give rise to this property 1. The following are examples of cell-based computation.

References

1

https://www.nature.com/articles/s41467-019-13232-z

2 Turing, A. M. On computable numbers, with an application to the Entscheidungsproblem.*Proc. Lond. Math. Soc.***s2–42**, 230–265 (1937).

3 Von Neumann, J. First draft of a report on the EDVAC.*IEEE Ann. Hist. Comput.***15**, 27–75 (1993).

4 Konkoli, Z. et al. Philosophy of computation. In*Computational Matter*(eds Stepney, S., Rasmussen, S. & Amos, M.), 153–184 (Springer, 2018).

5 Horsman, D., Kendon, V., Stepney, S. & Young, J. P.W. Abstraction and representation in living organisms: when does a biological system compute? In*Representation and Reality in Humans, Other Living Organisms and Intelligent Machines*, 91–116 (Springer, 2017).

6

https://www.sciencedirect.com/science/article/abs/pii/S0167779901016912

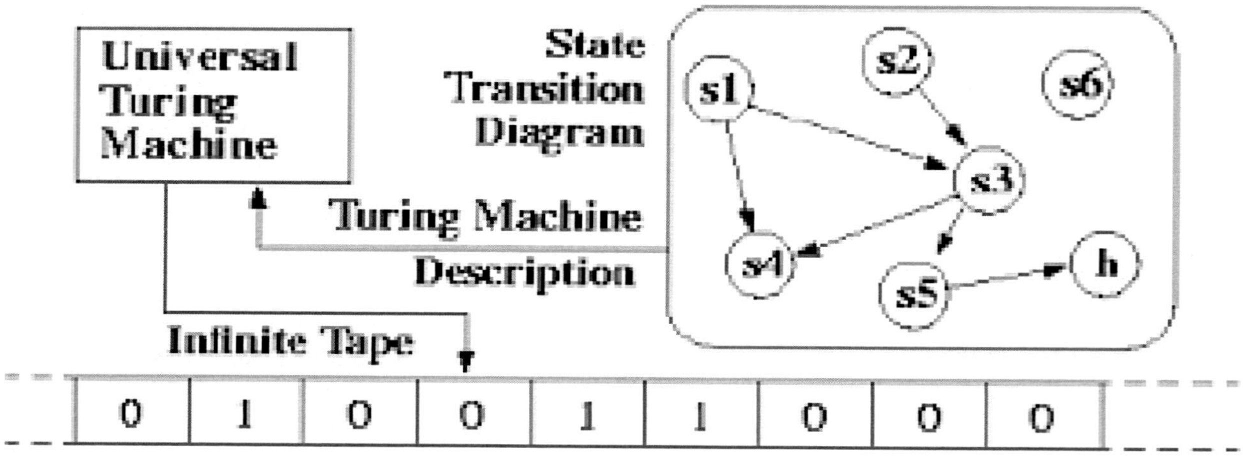

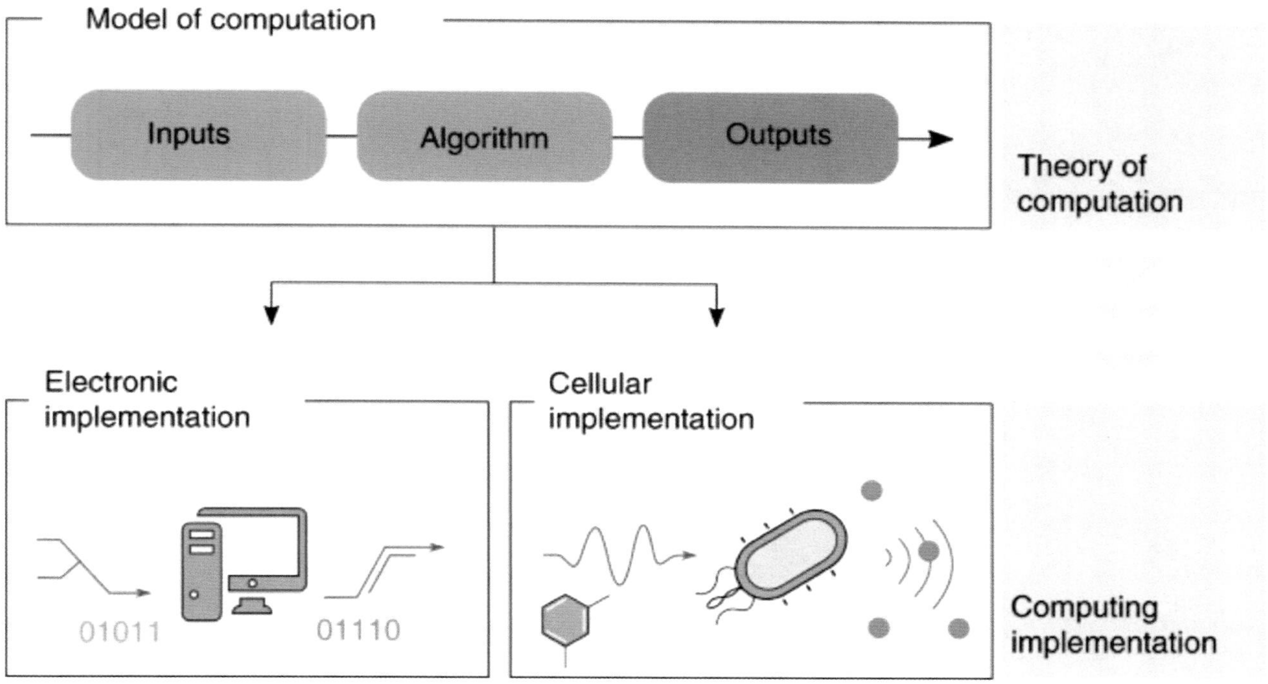

The cell as a "physical" computer. A model of computation formally defines inputs and outputs, as well as how an algorithm processes inputs into outputs. Though the same theoretical model of computation can be physically implemented in many different ways, the nature of computation remains the same 1.

Electronic implementations receive electronic data for inputs/outputs, while cells are able to sense/deliver a wide range of physical, chemical and biological inputs/outputs. The encoding of information into inputs can be done in different ways. Temperature, for instance, can be encoded as the height of mercury in a tube, the voltage of an electronic thermometer or the state of a DNA thermosensor 1.

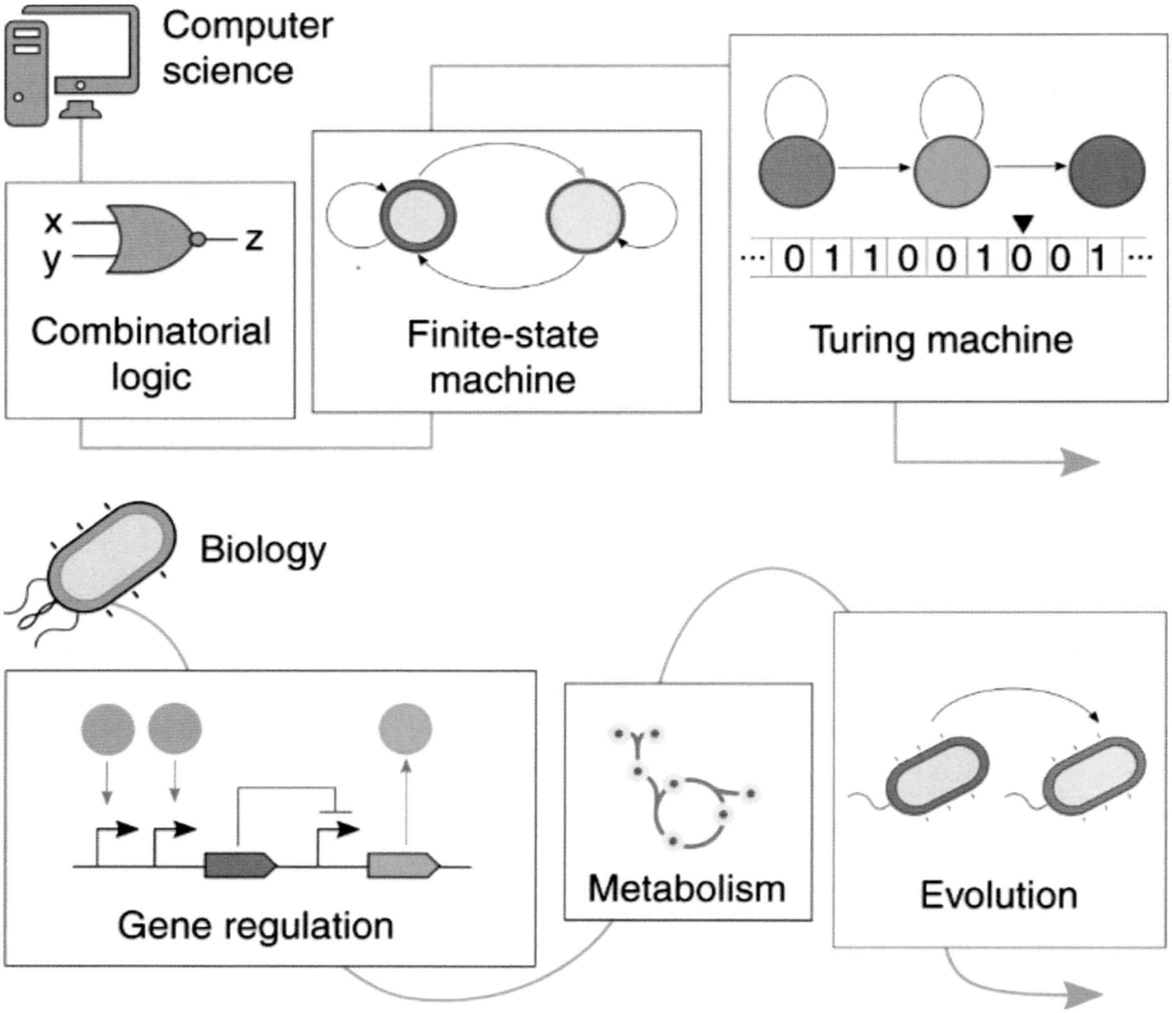

Mathematical models of computation and their properties inform the engineering of their physical manifestations . Many such implementations may be possible, but all inherit the characteristics of their abstract counterparts—both their abilities and their limitations. The fact that the nature of computation within a given model is independent of its implementation allows the application of theoretical computer science to all kinds of physical systems, including cells 1.

A variety of computational processes exist within cellular systems exceeding that of the Turing Machine. Biosensors represent only part of the picture of cellular activity. Individual cells or self-organized groups of cells perform extremely complex fuctions that include sensing, communication, navigation, cooperation and even fabrication of synthetic nanoscale materials. In natural systems, these capabilities are controlled by complex genetic regulatory circuits. Research efforts mimic the functionality of man-made information-processing systems within whole cells 6.

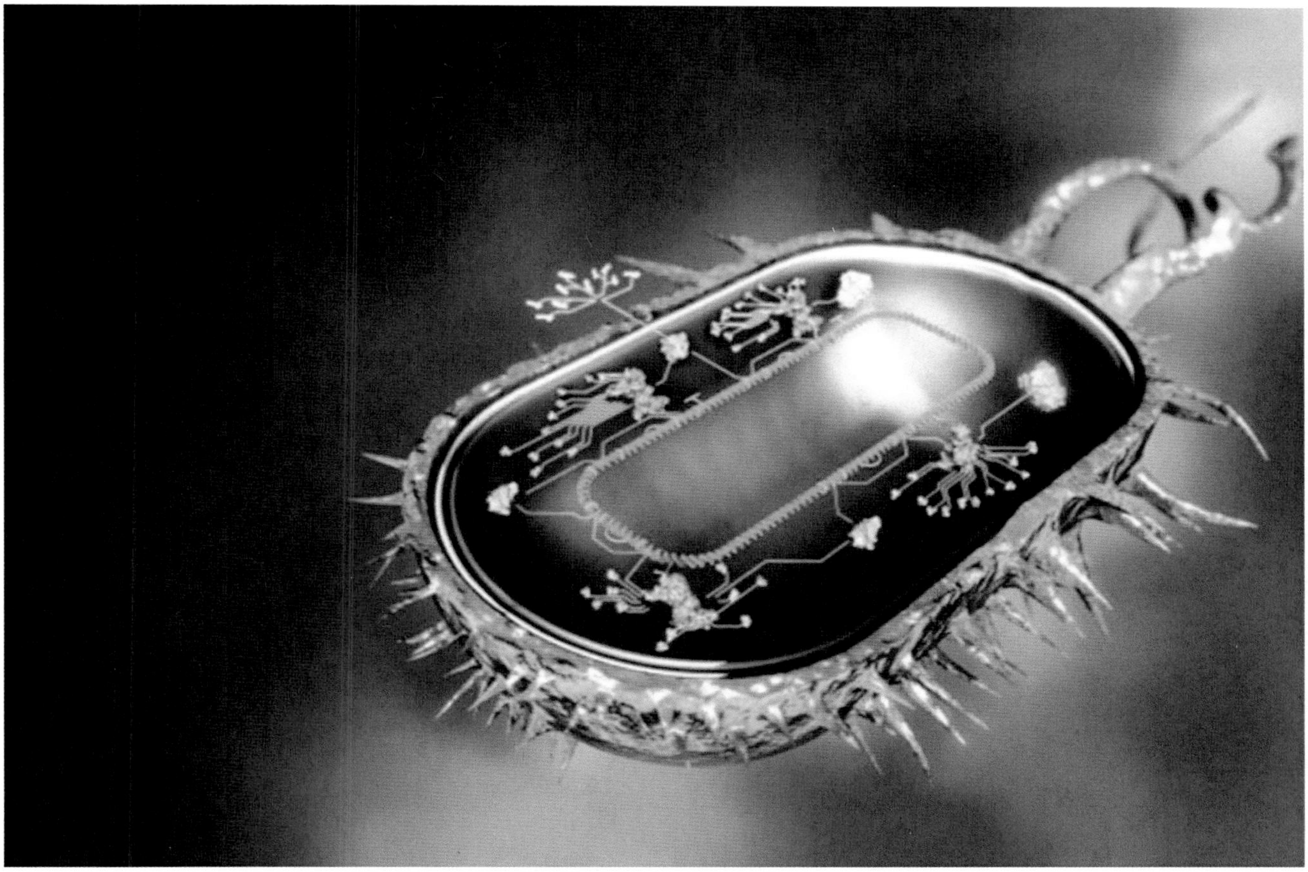

Information processing systems wthin biology are qualitatively different from those of classical computing. The overlaying of digital systems of logic is not easily accomplished within cells. This involves the mapping of sequential logic circuits onto seemingly equivalent genetic circuits. However, these same logic circuits are overly simplistic in comparison to the often hidden complexities of biological systems 1.

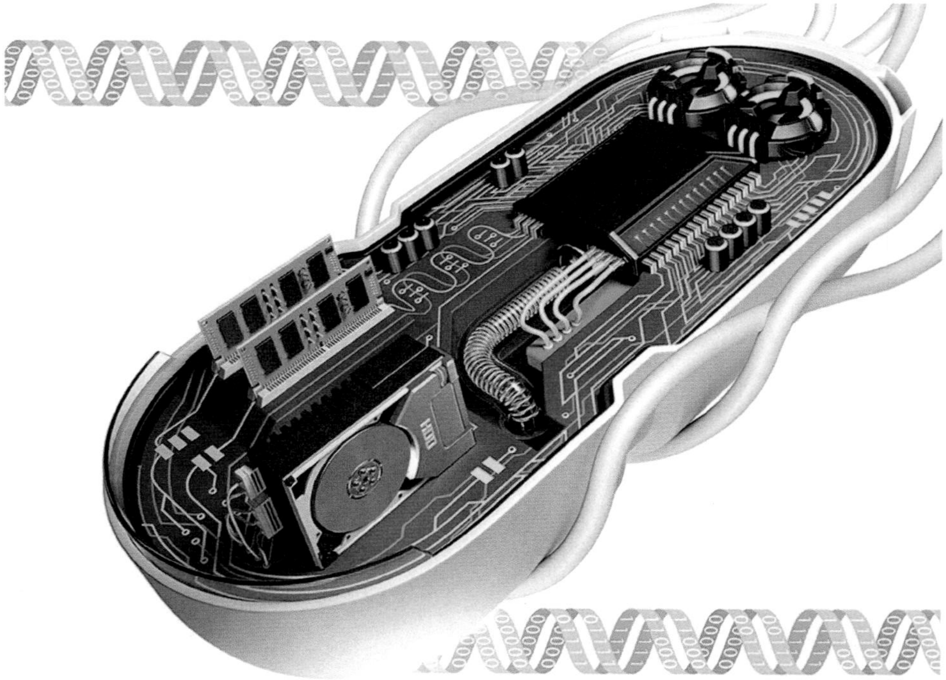

At the same time, designs for synthetic biological sensing and computing take inspiration from these same man-made systems. The relationship between computer science and natural computing is often synergistic. Yet, biological systems also consist of computational features that are unavailable to those based upon silicon 1.

For example, it has become increasingly clear that a number of biological processes show quantum mechanical properties [7,8]. In particular, there is strong experimental evidence that long-lived quantum coherence is involved in photosynthesis [9], and that quantum tunnelling is active in enzyme catalysis [10].

Evidence for quantum coherence

- Engel 2007: Quantum Beating: direct evidence of quantum coherence
- Lee 2007: "correlated protein environments preserve electronic coherence in photosynthetic complexes and allow the excitation to move coherently in space"
- Sarovar 2009: "a small amount of long-range and multipartite entanglement exists even at physiological temperatures."
- What does this mean for other biological systems?

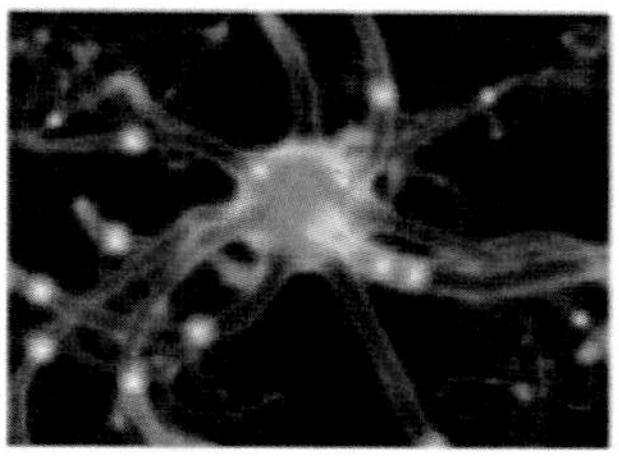

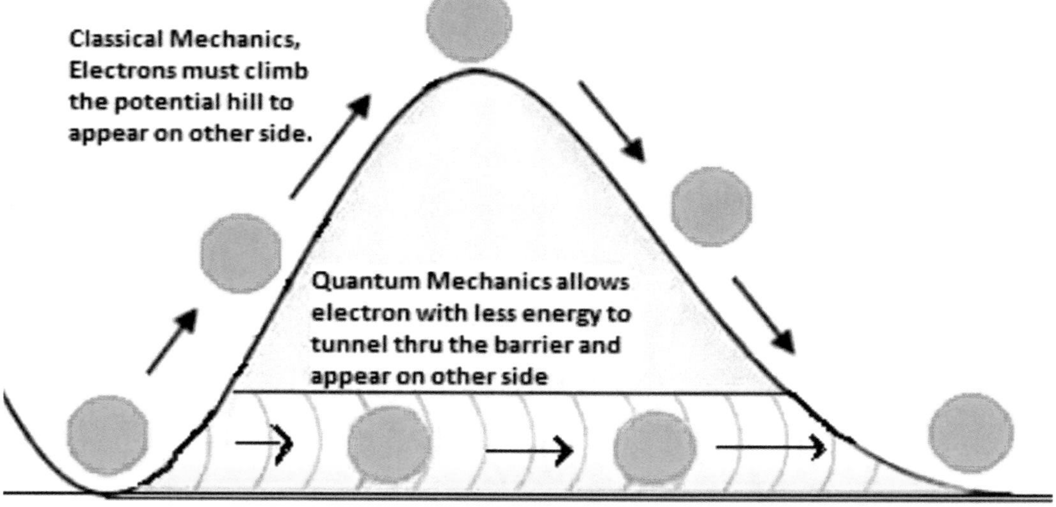

Since most quantum computing devices are built and run under stringent environmental conditions (at temperatures approaching absolute zero), the opportunity to control quantum effects in a biological system that "runs" at room temperature, through the emergence of a "quantum synthetic biology", could turn out to be a game changer in the quantum supremacy race. More to the point, realising models of quantum computation [11] using quantum biology could yield a cellular computer capable of a radically different kind of computation than silicon [1,9].

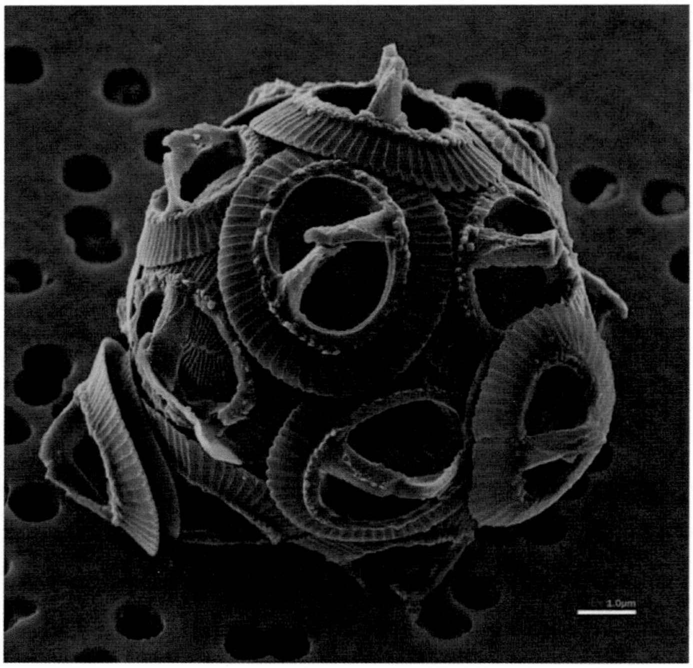

The more common forms of computation are digital; that is, information represented as a set of discrete values. Within electronic circuits, the continuous-value voltages (electron flow) must be broken into discrete forms representative of digital values (0,1). While analog computation allows for continuous signals to be used directly in the computational steps of an algorithm, and for represenation of information as continuous values [1].

References

1

https://www.nature.com/articles/s41467-019-13232-z

7 Lambert, N. et al. Quantum biology.*Nat. Phys.***9**, 10–18 (2013).

8 Adriana, M. et al. The future of quantum biology.*J. R. Soc. Interface***15**, 20180640 (2018).

9 Engel, G. S. et al. Evidence for wavelike energy transfer through quantum coherence in photosynthetic systems.*Nature***446**, 782–786 (2007).

10 R. K. Allemann & Scrutton, N.S.*Quantum Tunnelling in Enzyme-Catalysed Reactions*(Royal Society of Chemistry, 2009).

11 David, D. & Roger, P. Quantum theory, the Church–Turing principle and the universal quantum computer.*Proc. R. Soc. Lond. A. Math. Phys. Sci.***400**, 97–117 (1985).

12 Vladimirov, N. & Sourjik, V. Chemotaxis: How bacteria use memory.*Biol. Chem.***390**, 1097–1104 (2009).

13 Scialdone, A. et al.*Arabidopsis*plants perform arithmetic division to prevent starvation at night.*eLife***2**, e00669 (2013).

14 Sarpeshkar, R. Analog synthetic biology.*Philosophical transactions. Series A, Mathematical, physical, and Engineering sciences*, 372, 2014.

15 Woo, S. S., Kim, J. & Sarpeshkar, R. A digitally programmable cytomorphic chip for simulation of arbitrary biochemical reaction networks.*IEEE Trans. Biomed. Circuits Syst.***12**, 360–378 (2018).

Although many cellular computations involving binary "yes/no" decisions may be interpreted as digital computations, and digital logic computations are certainly suited to applications such as biosensors, cells often exhibit graded responses to stimuli that are more appropriately viewed as analogue computations [12,13]. Furthermore, the biochemical processes responsible for cellular computations involve discrete interactions of discrete molecules, but are also inherently stochastic. Cellular computing may, therefore, be viewed as both digital and stochastic, or as analogue computation with noise [14].

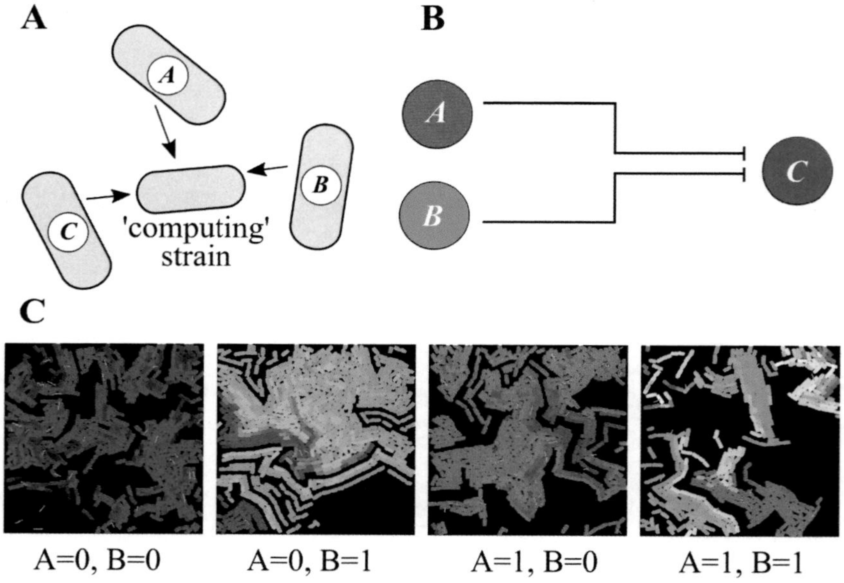

There are physical connections between chemistry and electronics [15], however, the cellular environment is a radically different computing substrate than silicon. Aside from gene regulation, which has been useful in engineering biological logic circuits, a number of processes and features exist in natural systems which offer computational capabilities. The following are four such resources [1].

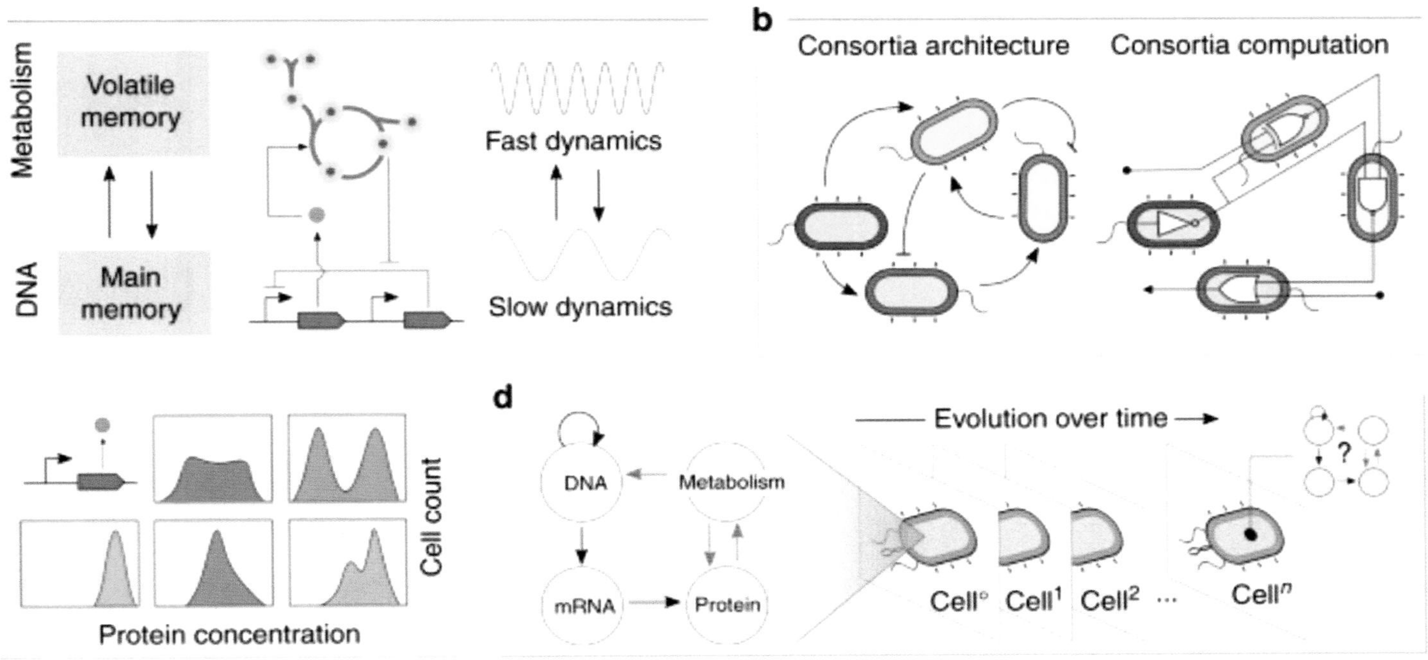

Cellular information-processing fundamentals that go beyond combinatorial logic circuits.

a Whole-cell computations, merging genetic and metabolic circuits, could achieve more ambitious goals than genetic circuits alone. Cells have evolved intricate networks that make simultaneous use of the varied features of both genetic and metabolic processes. In terms of information storage, metabolism presents a volatile memory, while DNA sequences are able to store information in a more stable fashion. Coordinating the use of different types of memory is a fundamental aspect of complex computer architectures.

The dynamic difference is also a potential source of complexity if coupled; metabolic reactions operate on a faster timescale relative to genetic regulatory networks 1.

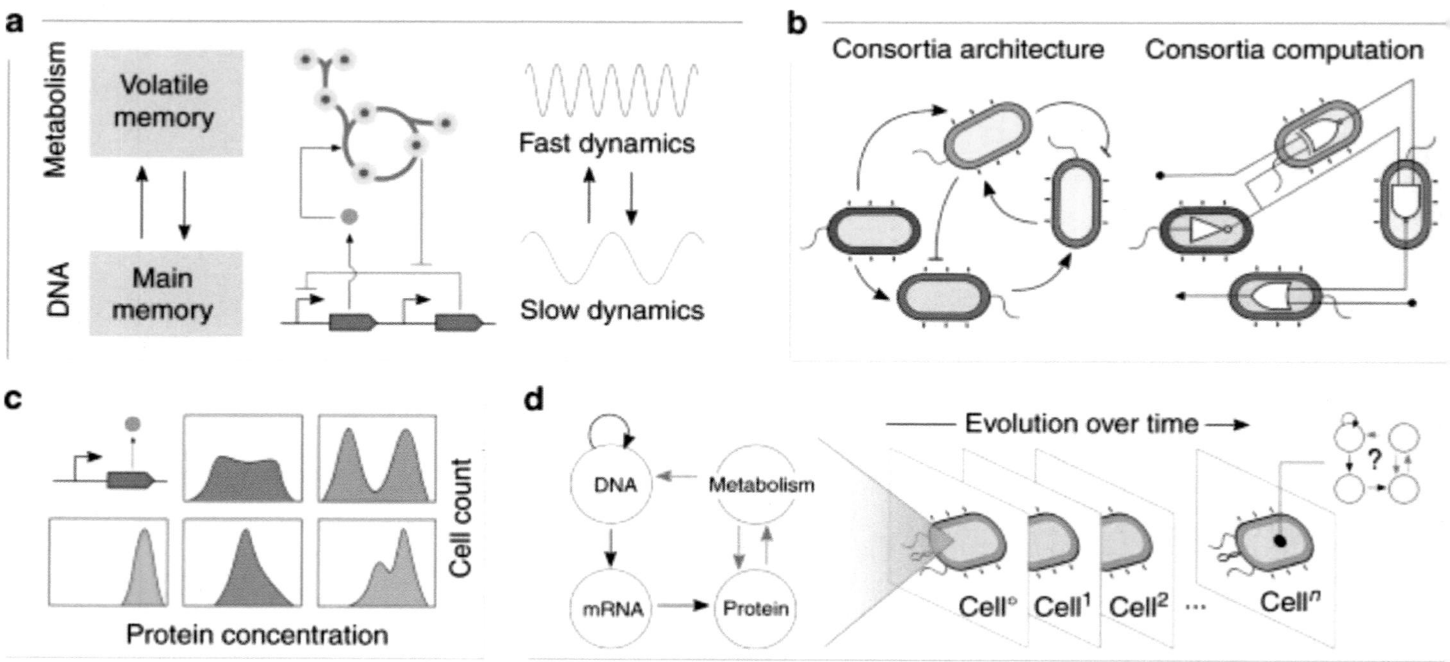

b Multicellular computing (right) is currently implemented by connecting the output of one strain to the input of another. Social interactions among cells (left), such as cooperation, mutualism, competition or commensalism, are not considered in general. However, social interactions are fundamental in natural communities—they provide stable architectures executing a desired computation 1.

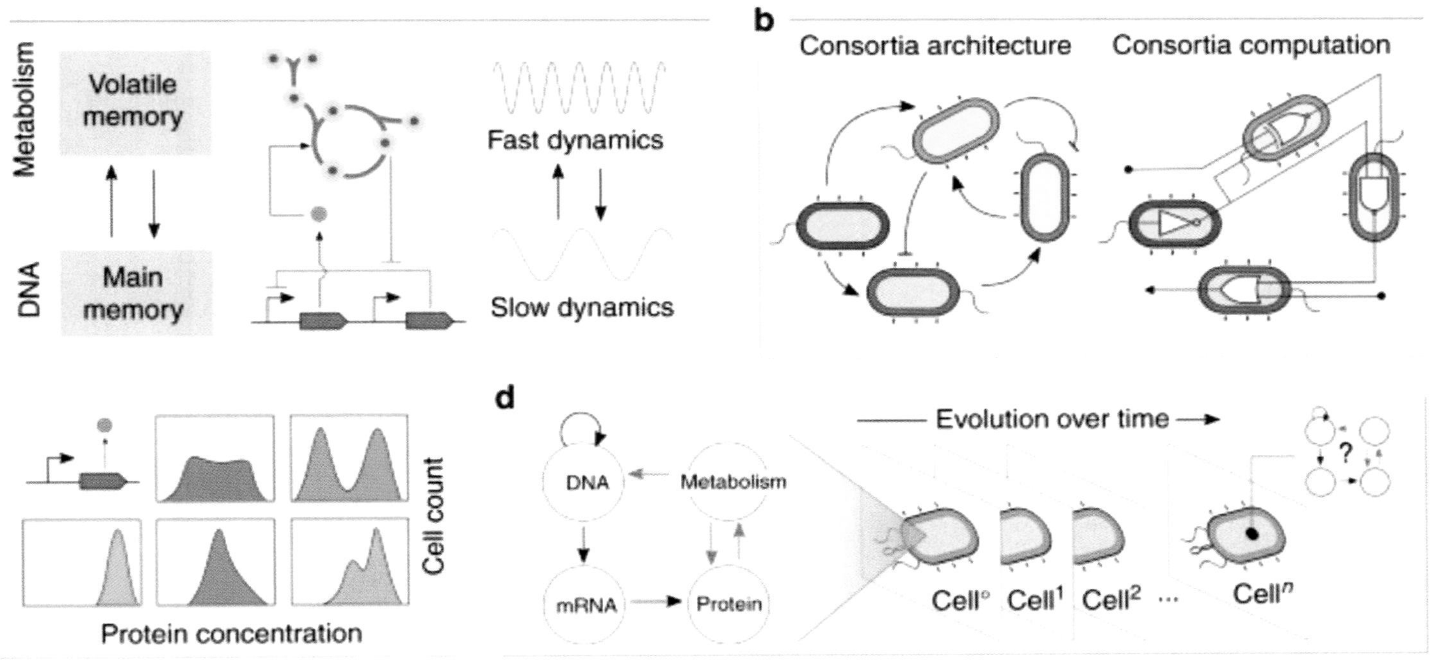

c Gene expression noise is intrinsic to living systems; the panel figure shows different patterns for gene expression. Despite the fact that all are described as being on, there are different types of expression—thus different on/off standards 1.

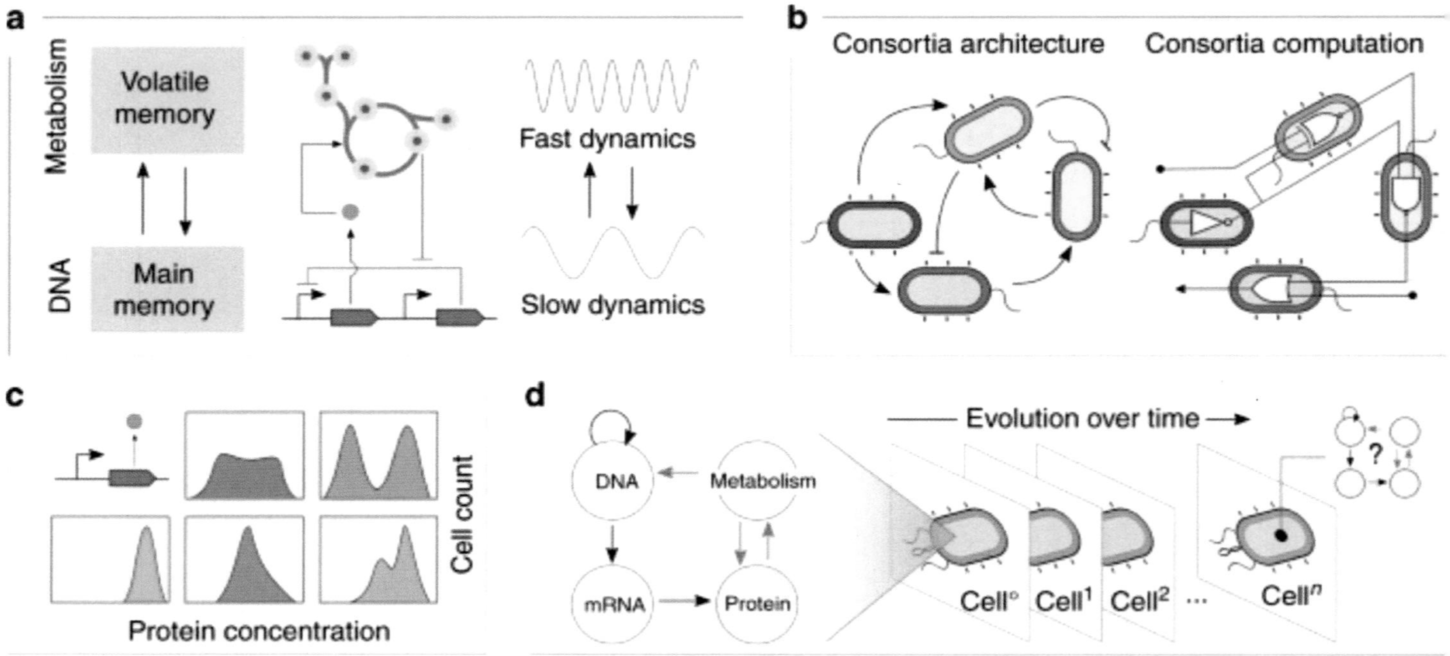

d The cell as a general-purpose machine. As the basis for a model of computation, the central dogma of molecular biology can be expanded to include metabolism. Evolutionary processes may also be included as major forces guiding information-processing in cells, since they allow the purpose of cellular computations to adapt over time 1.

Natural cellular computing operates at vast scales, in a distributed manner, and in the presence of considerable noise [1] (stochastic). Consequently, biological metaphors have served as inspirations for models of amorphous computation [1].

Robotics has also drawn inspiration from biological computation, particularly in relation to morphological computing, which takes advantage of the physical properties of computing agents in order to achieve more efficient computations [1,16].

In the context of embodied artificial intelligence, morphological computation refers to processes, which are conducted by the body (and environment) that otherwise would have to be performed by the brain.

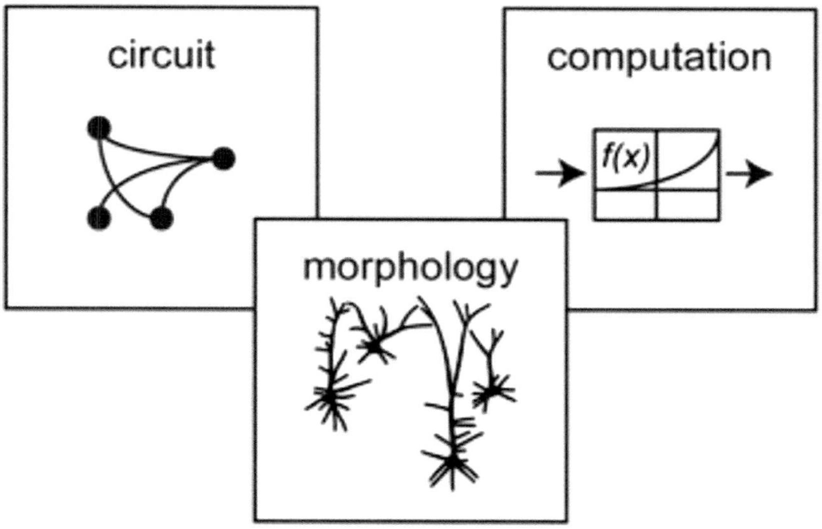

By using intrinsic physical properties of the computational substrate to "outsource" parts of the computation, increasingly complex computations can be carried out while maintaining relatively simplistic control structures [1,17].

Living systems are an ideal implementation technology for morphological computation (biological systems using their bodies to control basic actions) since they not only (naturally) compute solutions to individual instances of problems, but continuously compute and adapt in order to embody an efficient general solution. Conventional silicon computers have inflexible architectures by comparison, which must sacrifice efficiency for generality. These qualitative differences between cellular and conventional computing suggest that applications such as terraforming and smart material production may remain beyond the reach of silicon computers, but in contrast, strategies for both applications based on living technologies have already been proposed [1,18,19].

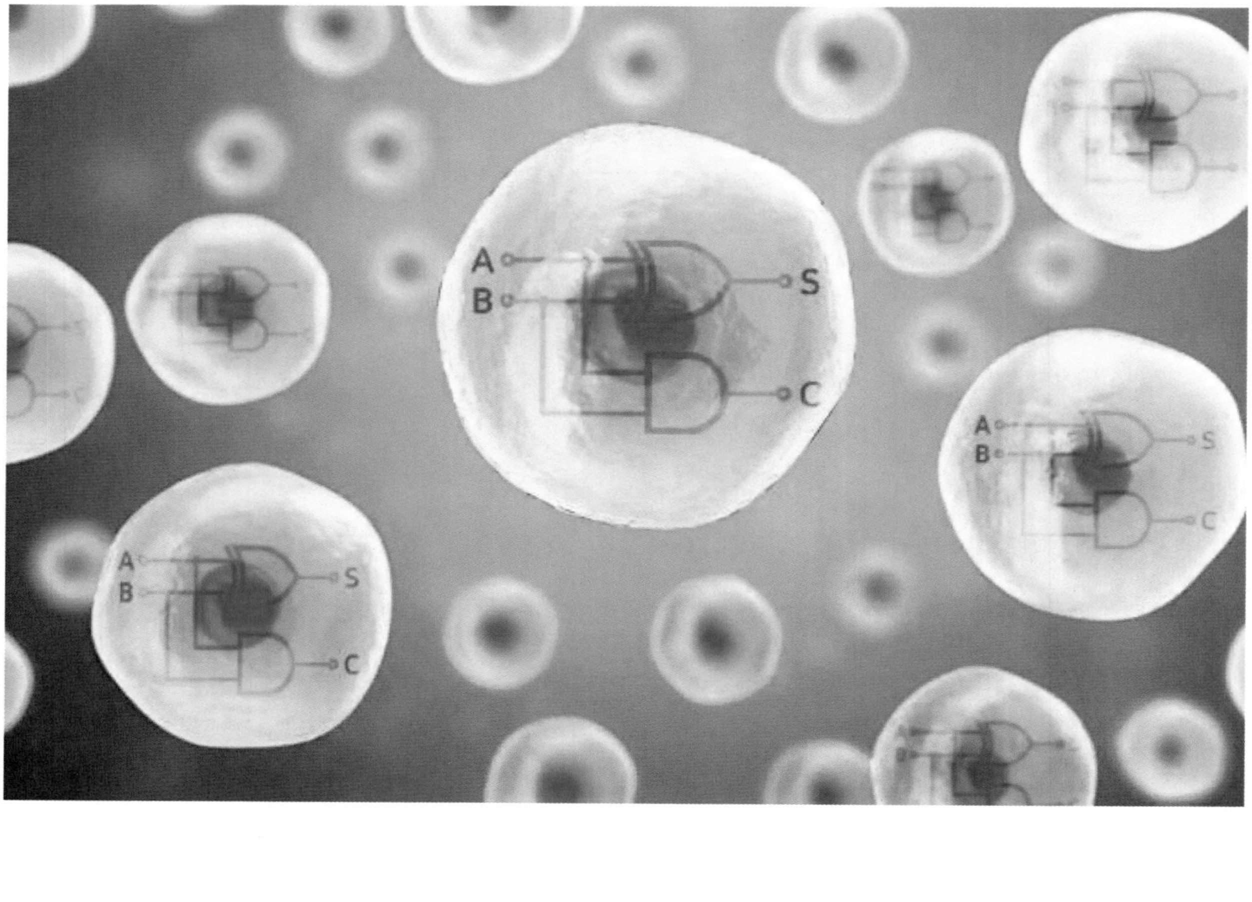

References

1

https://www.nature.com/articles/s41467-019-13232-z

16 Gordana D.-C. The info-computational nature of morphological computing (ed. Müller, V. C.),*Philosophy and Theory of Artificial Intelligence*, Studies in Applied Philosophy, Epistemology and Rational Ethics, 59–68 (Springer Berlin Heidelberg, Berlin, Heidelberg, 2013).

17 Umedachi, T., Takeda, K., Nakagaki, T., Kobayashi, R. & Ishiguro, A. Fully decentralized control of a soft-bodied robot inspired by true slime mold.*Biol. Cybern.* **102**, 261–269 (2010).

18 Solé, R. Bioengineering the biosphere?*Ecol. Complex.* **22**, 40–49 (2015).

19 Armstrong, R. Systems architecture: A new model for sustainability and the built environment using nanotechnology, biotechnology, information technology, and cognitive science with living technology.*Artif. Life* **16**, 73–87 (2010).

20 https://www.ncbi.nlm.nih.gov/pmc/articles/PMC6462112/

21 Mammalian designer cells: Engineering principles and biomedical applications.

Xie M, Fussenegger M

MICROSOFT, 2016: "WE CAN PROGRAM COMPLEX BEHAVIORS USING DNA". 3-STRAND DNA CONFIRMED

by Silviu "Silview" Costinescu March 1, 2021

Excerpts:

To me, the most striking part in this video is the confirmation that they are after the three-stranded DNA technology Anthony Patch brought up in that sensational 2014 interview, which also earned us a ban from Youtube.

"Imagine a biological computer that operates inside a living cell"
– Dr.Andrew Phillips, head of bio-computation at Microsoft Research.

"The problem we're trying to solve is really trying to have a more sophisticated diagnosis that can happen automatically inside cells... In this project, we're trying to use DNA as a programmable material" according to Dr.Neil Dalchau, a scientist at Microsoft Research.

"[Microsoft] are essentially trying to sense, analyze andcontrol molecular information"

Georg Seelig, Associate Professor at theGates-funded University of Washington.

Moderna described mRNA as “aninformation molecule” and even trademarked the name “mRNA OS” – meaning ‘operating system’, according tobigtechtopia.com We have Moderna’s head honcho “on tape” describing the mRNA vaccine as “information therapy”:

“Molecular devices made of nucleic acids show great potential for applications ranging from bio-sensing to intelligent nanomedicine. They allow computation to be performed at the molecular scale, while also interfacing directly with the molecular components of living systems. They form structures that are stable inside cells, and their interactions can be precisely controlled by modifying their nucleotide sequences. However, designing correct and robust nucleic acid devices is a major challenge, due to high system complexity and the potential for unwanted interference between molecules in the system.

To help address these challenges we have developed the DNA Strand Displacement (DSD) tool, a programming language for designing and simulating computational devices made of DNA. The language uses DNA strand displacement as the main computational mechanism, which allows devices to be designed solely in terms of nucleic acids. DSD is a first step towards the development of design and analysis tools for DNA strand displacement, and complements the emergence of novel implementation strategies for DNA computing."

Microsoft Research

For more information, please visit: https://www.anthonypatch.com

Made in the USA
Middletown, DE
24 September 2021

48995197R00020